Historic England

AF879596

Ships and Boats: Prehistory to Present

Designation Selection Guide

Summary

Historic England's selection guides help to define which historic assets are likely to meet the relevant tests for national designation and be included on the National Heritage List for England. The 2002 National Heritage Act expressly gave Historic England new responsibilities for the preservation of monuments in, on or under the seabed. The Council of Europe's 1992 European Convention on the Protection of the Archaeological Heritage (known as the Valetta Convention of 1992, and ratified by the UK in 2001) makes no distinction between land and sea in terms of the location of archaeological assets. Protecting wreck sites is part of our duty to preserve this aspect of the marine historic resource. The existing arrangements are complex, so one of the principal aims of this selection guide is to set out current approaches and provide extra clarity as to our designation approaches.

This guide applies to all vessels, irrespective of the original environment they navigated, covering those used on inland waters, coastal waters and the open sea. It also includes vessels that are now buried under the ground, where reclamation or some other process has caused a former waterway to be covered by dry land: many of the most significant early boats and ships have been discovered on land rather than at sea. It also deals with vessels that are no longer afloat, and which have become permanently settled (either on land, in an inter-tidal location, or within English territorial waters, which extend up to 12 nautical miles from the coast), and which meet the legal test of forming a vessel or its remains.

This guidance note has been prepared by the Listing Team.

First published by English Heritage May 2012.

This edition published by Historic England September 2016.

HistoricEngland.org.uk/advice/

Front cover
View of two post-medieval beached vessels, Elmley Ferry, Isle of Sheppey (Kent).

Contents

Introduction and Definitions 1

Scope of selection guide 2
The present state of designation 3
Other forms of recognition and protection 4

1 Historical Summary 5

1.1 Ships and boats 5
1.2 Marine archaeology 9
1.3 Understanding and quantifying the resource 10

2 Special Considerations 13

2.1 Listing 13
2.2 Scheduling..14
2.3 Protected Wreck Sites..............................17

3 Where to Get Advice...................18

3.1 The National Register of Historic Vessels..18
3.2 Military maritime graves...........................18
3.3 Designation of the natural environment..19
3.4 Contact Historic England 20

4 Select Bibliography 21

4.1 General 21
4.2 Prehistoric and Roman 21
4.3 Medieval 21
4.4 Post-Medieval...21
4.5 Websites 22

5 Acknowledgements 23

Introduction and Definitions

The receding sheets of the last Ice Age shaped the British Isles about 13,000 years ago, and eventually severed our physical links with the Continent between about 6500 and 6200 BC. Since then, our national story has been inextricably linked to the surrounding seas: boats and ships have determined the course of much of our history – through conquest, defence, migration and commerce. They have played a fundamental role in the emergence of Britain as a world power, which was firmly founded on maritime might as well as on industrial pre-eminence. How we protect their remains is a topic of increasing interest (Fig 1).

However, some will ask why any vessels should be designated and protected through the planning and heritage protection systems. Is not their investigation prohibitively costly? Isn't protecting ships best left to maritime museums? And who can oversee due control of protected areas? The 2002 National Heritage Act expressly gave Historic England new responsibilities for the preservation of monuments in, on or under the seabed. The Council of Europe's 1992 European Convention on the Protection of the Archaeological Heritage (known as the Valetta Convention of 1992, and ratified by the UK in 2001) makes no distinction between land and sea in terms of the location of archaeological assets. Protecting wreck sites is part of our duty to preserve this aspect of the marine resource (Fig 2).

Figure 1
Protected Wreck Site marker buoy.

Marine protection is warranted for several reasons. Firstly, the sites in question can yield much information about ships, mercantile trade, the lives of sailors and passengers and about society in general. Secondly, owing to developments in navigation and location-finding, it is now possible to identify with great precision the location of identified wrecks and ensure that due allowance is given to their survival. Thirdly, some modern practices (such as dredging, gravel extraction, trawling, wind farm construction and pipe-laying) are so destructive of the seabed

Figure 2
Diver Survey, St *Antony* Protected Wreck Site, Cornwall.

and its deposits that identification is essential if certain zones of proven sensitivity are to be safeguarded. Fourthly, marine planning is rapidly developing and is based on an ecosystems approach which recognises that the protection and management of marine cultural heritage is based on an understanding of its significance. Fifth and finally, designation can assist in widening the already-considerable public appreciation of England's maritime heritage.

How we assign significance to vessels (understood here as being simply a general term to describe all kinds of craft designed for transportation on water) warrants explanation. The existing arrangements are complex, so one of the principal aims of this selection guide is to set out current approaches and provide extra clarity as to our designation approaches. An increasing amount is becoming known about historic vessels and Historic England is committed to the development of appropriate protection responses.

Scope of selection guide

This selection guide applies to all vessels, irrespective of the original environment they navigated: it covers vessels used on inland waters, coastal waters and the open sea (Fig 3). It also includes vessels that are now buried under the ground, where reclamation or some other process has caused a former waterway to be covered by dry land: many of the most significant early boats and ships have been discovered on land rather than at sea. It deals with vessels that are no longer afloat, and which have become permanently settled (either on land, in an inter-tidal location, or within English territorial waters, which extend up to 12 nautical miles [a nautical mile is approximately 1.15 statute miles, or 1.85 km] out from the coast), and which meet the legal test of forming a vessel or its remains.

This Selection Guide also includes a consideration of protection approaches to other forms of vehicles and aircraft which may be encountered in a maritime archaeological context. It does not cover vessels, vehicles or aircraft which are still mobile, although the scope of the National Register of Historic Vessels, the National Historic Fleet, and The Protection of Military Remains Act 1986 are outlined for clarity's sake (below, section 2).

As its title suggests, this selection guide concentrates on ships and boats. It does not, therefore, cover the equally significant realm of other archaeological sites which lie beneath the sea. These include those prehistoric sites which lie on Doggerland, the landscape between Britain and the Continent lost to rising sea levels between about 6500 and 6200 BC. These are considered in the Settlement Sites scheduling selection guide. Further guidance on our designation approaches to submerged sites will be produced as and when appropriate.

A very brief historical overview on the development of vessels is included below (Section 1). A longer treatment is given in Historic England's 'Introductions to Heritage Assets', Ships and Boats: Prehistory to 1840 and Ships and Boats 1850 to 1950.

Figure 3
Modern metal hulk, Minster Beach, Isle of Sheppey, Kent.

The present state of designation

As this Selection Guide sets out in below, in Section 2, a number of designation approaches have been deployed in the field of ships and boats. With the exception of the National Register of Historic Vessels, these approaches are overseen by Historic England which makes its recommendations to the Secretary of State for Culture, Media and Sport for final determination.

In 1973, the passing of the Protection of Wrecks Act enabled the statutory recognition of important wreck sites to take place, with measures put in place for their safeguarding through the granting of licences for their oversight, and supervision of access: to date, fewer than 60 sites in English waters have been designated as **Protected Wreck Sites**. This is a very small proportion of documented wreck sites and known ship losses, which currently number around 37,000, which is itself only a tiny proportion of all vessels lost or left to decay over the centuries.

Hitherto, **scheduling** has been sparingly deployed; although the Ancient Monuments and Archaeological Areas Act specifically permits the scheduling of any site comprising a vessel or remains of a vessel, fewer than a dozen items among the 20,000-odd entries on the schedule include vessels. Scheduling has rarely applied to ships and boats, including some inter-tidal sites where the full benefits of the management options offered by scheduling can be realised.

Listing, which identifies buildings and structures of architectural or historic interest, has also been very occasionally applied to a few permanently secured vessels such as the Grade I *Cutty Sark* in Greenwich, as well as to structures incorporating nautical elements.

In overall numbers, therefore, the designation tally for historic vessels is very small. As our appreciation of the significance of maritime heritage develops, further additions to the National Heritage List for England will be made as appropriate.

Other forms of recognition and protection

As set out below in section 3, there are other means of affording recognition, and in some cases protection, to the sites of historic vessels beside those of statutory heritage designation. These include natural designations, such as Sites of Special Scientific Interest (SSSI) and Ramsar Sites (wetland sites protected under the terms of the Ramsar Convention of 1971), and Marine Protected Areas which are focussed on protecting biodiversity in inter-tidal and sub-tidal zones. Government guidance, headed by the UK Marine Policy Statement (2011) and the National Planning Policy Framework (2012), stresses the importance of according **all** heritage assets of significance due consideration, irrespective of their formal designation status. Historic England's preference is that wherever possible, other existing systems of protection should be used to champion heritage protection considerations in order to make regulation as efficient and as stream-lined as possible.

With regard to heritage protection and marine planning, Historic England has worked with the Department for Environment, Food and Rural Affairs (Defra) to deliver the marine planning system set out in the Marine and Coastal Access Act 2009; a reformed system of marine development licensing; the identification of Marine Conservation Zones; and the integration of cultural heritage within the management of inshore fisheries within 6 nautical miles (nm).

Substantially intact vessels which have demonstrable and significant associations with the UK may be added to the National Register of Historic Vessels (NRHV), which is administered by the organisation National Historic Ships UK. These include both vessels afloat, and vessels on permanent display in museums, and as a consequence the NRHV forms a very important parallel system to that of heritage designation (Fig 4).

In 2009 the UNESCO Convention on the Protection of the Underwater Cultural Heritage entered into force and is intended to enable states to better protect their submerged cultural heritage. The Convention sets out basic principles for the protection of underwater cultural heritage and consists of a main text and an annex, which sets out the 'Rules for activities directed at underwater cultural heritage'. Although the UK has not ratified the Convention, the 'Rules' of the convention were adopted by the UK in 2005 as being best practice for underwater archaeology and are a further step towards the safeguarding of underwater cultural heritage.

More information is provided on approaches to designation in Sections 2 and 3 of this selection guide.

Figure 4
Paddle Steamer *Ryde*, Isle of Wight, launched 1937, part of the UK's National Historic Fleet.

1 Historical Summary

1.1 Ships and boats

The evidence for water transport in England begins with a few finds of Neolithic canoes dating from around 4000 BC. By the Middle Bronze Age seagoing stitched-planked boats (that is, where a vessel's planks abut, rather than overlap, and are joined by hide or bark thread stitched through holes) were in use. Three have been found: the seventh or eighth century BC raft at Brigg in Lincolnshire; the plank boat from about 2000 BC at Ferriby on the River Humber (Yorkshire); and the Dover Boat in Kent dated to 1550 BC – the earliest known examples in the world. The potential for the discovery of further examples of early craft – in this case in secure, stratified, contexts – was demonstrated in 2011 at Must Farm, near Peterborough (Cambridgeshire), when six dugouts, provisionally dated to the Middle Bronze Age, were discovered in a clay-pit. Artefacts (such as pottery, metal and other portable goods) also tell of extensive, and complex, pre-Roman sea and river trade routes (Fig 5).

Roman written sources record a wide range of ocean-going vessels - native as well as Roman – and the remains of some have been found in London and elsewhere, some as wrecks and others in other contexts, such as a second-century AD scheduled example at New Guy's

Figure 5
The Iron Age logboat from Poole, Dorset.

Figure 6
Bow timbers of the New Guy's House Roman Boat, Southwark. Discovered 1959, Scheduled 1983.

House, Bermondsey (Fig 6). In the post-Roman period new types of boats and ships were developed in north-west Europe including the early seventh-century forty-oared ***ceol*** (keel) found near Ipswich (Suffolk) in 1939 known as the Sutton Hoo ship; together with the royal burial deposit it contained, this is perhaps one of the most outstanding archaeological discoveries ever made in England. It has many similarities with the Viking longships – none of which have been found in England, although several ship burials have been found in Scotland – which raided and traded across Europe and beyond from Scandinavia for 300 years. The first documentary evidence of a purpose-built English royal naval force comes from the time of Alfred (King of Wessex 871-99), when a fleet of large, oared, ships was built to help fight the invading Danes. Locally far more modest craft including dugouts, punts and skin boats continued in use as ferries and for carrying local trade and fishing around England's coast and on its extensive network of rivers and man-made waterways, such as the ninth to tenth-century AD Anglo-Saxon boat discovered at Graveney, Kent, in 1970.

In the twelfth and thirteenth centuries larger and heavier vessels including cogs and hulks facilitated ever bulkier and heavier trade on European seas with extensive networks extending across the continent. The evidence for England however is primarily documentary and iconographic (although the thirteenth-century Magor and fifteenth-century Newport ships, found in South Wales, probably traded in English ports around the south-west, as well as with north-western France and Spain). Carracks, large merchant ships with rounded hulls and fore-and aft-castles, were being built by the end of the fourteenth century. English clinker-built carrack-type vessels were probably first constructed in 1413-18 with the four 'great ships' of Henry V, the fourth being the *Grâce Dieu*, launched about 1418 whose remains lie in the River Hamble near Southampton, Hampshire, near to those identified as the early fifteenth-century warship *Holigost* (Fig 7). Perhaps the most important ship ever to be lifted from English waters is the *Mary Rose*, now displayed in Portsmouth Dockyard, Hampshire. Constructed on Henry VIII's orders in 1509-11,

Figure 7
Henry V's *Grâce Dieu* about 1875 on the River Hamble, near Bursledon, Hampshire; designated a Protected Wreck Site in 1974.

Figure 8
Seventeenth-century galley bricks from the Norman's Bay Protected Wreck Site (East Sussex), a large warship of the period 1600-1800.

Figure 9
The late nineteenth-century Wheel Wreck, Protected Wreck Site, Isles of Scilly.

the *Mary Rose* sank in the Solent in combat in 1545, and is a remarkable testament to the Tudor expansion of the Royal Navy, which also saw the opening of a network of royal dockyards.

Ever larger ships were commissioned by the Navy in the seventeenth and eighteenth centuries (Fig 8). The East India Company (founded in 1600) brought similarly ambitious ships into service, and overall the total tonnage of the English merchant fleet rose at least threefold between the early 1700s and the late 1780s; by 1800 the British merchant fleet was the largest in the world. Those ships included slavers involved in the notorious Atlantic slave trade, the abolition of which was – in British ships at least – finally enacted by Parliament in 1807. The age of wooden warships reached a peak in the late eighteenth and early nineteenth centuries, culminating in the French Revolutionary and Napoleonic Wars (1793-1815), after which many fighting ships were laid up or scrapped. Throughout this period a large number of these vessels were lost in naval or commercial service in English waters, from the once-common collier brig type represented by the eighteenth century Seaton Carew wreck site to examples of warships and commercial vessels, including the East Indiaman *Admiral Gardner*, which foundered on the Goodwin Sands off Kent in 1809, and is now a Protected Wreck Site.

While ships of iron and with screw propellers had been built before, Brunel's *SS Great Britain*, launched in 1843, was the first to combine these in a large, ocean-going, ship, intended for service between Bristol and New York. Steam, introduced to paddle-driven steam auxiliaries from the 1820s, came to the battle fleet in 1845 with propeller propulsion, and over the following years a revolution in naval vessels took place as iron plating and breech-loading guns firing high-explosive shells rendered the old wooden ships of the line obsolete. Over the next half-century, warships generally became heavier and ever-bigger gunned, leading to the construction of HMS *Dreadnought* in 1906: a fast, turbine-driven battleship, with massive firepower. Naval vessels were designed to incorporate new developments in ordnance, such as torpedoes and depth charges, which developed rapidly during the First World War. The arrival of the submarine in the early twentieth century would have major strategic and tactical implications for the Royal Navy's capital ships too.

Merchant shipping followed a similar development trajectory culminating in the great transatlantic liners of the early twentieth century, although modernisation of propulsion machinery was largely confined to the adoption of the steam turbine engine (Fig 9). Traditional sailing vessels remained in service in both local coasting and international routes where coal supplies were hard to come by, but were becoming ever more obsolete; the First World War hastened their demise. The war saw radical developments in both warships and mercantile vessels, including the pioneering use of innovative materials, such as ferro-concrete (steel-reinforced concrete) instead of more traditional materials such as steel or wood for ship construction, and new methods of shipbuilding: the British 'War Standard Ships' of both World Wars and the American 'Liberty Ships' made use of standardised design and prefabrication respectively, meaning that new ships could be turned out very quickly to replace tonnage sunk. Steam remained the principal means of propulsion, but the inter-war period saw the development of motor engines retrofitted onto relatively small sailing vessels or designed into luxury leisure craft. Another characteristic of the period was the increasing size and tonnage of both warships and merchant vessels, which continued following the Second World War as modern ships increasingly turned towards the use of oil as fuel. The demand for petroleum products in itself drove up the size of vessels, as did containerisation, which permitted ever-greater bulk cargoes to be carried on a single vessel, and had a major impact on the design of ports.

The Second World War witnessed several major technological shifts in specifically military craft, with large-calibre battleships and cruisers predominant in the early stages of the war (a surviving example being the light cruiser HMS *Belfast* (commissioned 1939), on permanent display on the Thames in central London), before the rise of the central role of aircraft carriers later in the war as the pre-eminent 'capital' ship in the late war and post-war periods.

The later twentieth century saw a downward trend in the use of traditional vessel types as containerisation and inland circulation by road freight became the preferred forms of transport, thereby rendering barges and other vessels effectively obsolete in preference to speed and convenience (Fig 10). However, while commercial

Figure 10
The barge *Harriett*, abandoned at Purton, Gloucestershire. Scheduled in 2010.

Figure 11
The stern of Holland No. 5 submarine sunk on the Royal Sovereign Bank in the English Channel in 1912. Designated a Protected Wreck Site in 2005.

vessels became ever larger, and traditional fishing vessels were replaced by factory ships (also known as fish-processing vessels, large ocean-going vessels with extensive on-board facilities for processing and freezing caught fish or whales), there was a shift towards looking upon the sea as a leisure, rather than commercial, resource, so that the trend towards recreational craft, already seen in the late nineteenth century, became increasingly important and became a vehicle for further innovations in ship construction, such as the use of carbon fibre in yacht design.

The first, experimental, English submarines were built in the 1620s. It was only at the very end of the nineteenth century that militarily effective submarines began to be developed as electric and diesel propulsion, periscopes and torpedoes became standard features (Fig 11). The role played by submarines in both World Wars was of considerable strategic importance. During the First World War, unrestricted submarine warfare (including mine-laying) had a considerable impact upon the import of foodstuffs and military equipment into the UK. During the Second World War the impact of submarines was even more pronounced, leading to punishing losses of merchant ships by the allies that nearly led to defeat during the early stages of the war, at one point virtually stopping the import of materials into the UK. The submarine war only began to turn in the allies favour after the full use of the convoy system and introduction of better escort vessels (especially aircraft carriers and other technological innovations including better long-range detection of submarines) used alongside intelligence from code-breaking activities from mid-1943 onwards. Nuclear-powered, and in time nuclear-armed, vessels entered service in the 1950s, and equipment was introduced which extracted oxygen from seawater; together these greatly extended the range and capabilities of submarines, turning them into the most potent of all naval weapons during the Cold War.

1.2 Marine archaeology

Until the mid twentieth century, it was the desire to recover salvage – cargoes, and recoverable fabric – which drove the invention of devices which allowed men to work underwater. In 1690, the astronomer Edmond Halley invented a submersible diving bell with a renewable air supply. In 1715 John Lethbridge's pressurized diving suit enabled divers to spend an hour submerged. Then, in 1823, Charles and John Deane of Deptford invented a 'smoke helmet' breathing apparatus to allow breathing in fires, which they soon adapted for diving use, taking on contracts in the 1830s from the Admiralty to salvage wrecks. This, an 'India-rubber dress ... the head guarded by a metal helmet, having plate-glass in front' was effectively the closed-helmet diving suit, as used well into the twentieth century. Siebe Gorman and Company Ltd simultaneously developed diving equipment and advertised themselves as 'submarine engineers'. The company was founded by Augustus Siebe (1788-1872), a German-born British engineer.

Figure 12
Historic England marine archaeologist working in the Thames Estuary.

Figure 13
Diver Survey, Salcombe Cannon Protected Wreck Site, Devon.

A number of ships and boats were uncovered during harbour works from the early nineteenth century onwards. Excavation of burial mounds in Scandinavia uncovered Viking longboats, broadly comparable to that found at Sutton Hoo in 1939. Marine or underwater archaeology in the sense of the investigation and recovery of ships, like the *Vasa*, lost in 1628 off Stockholm (Sweden) and raised largely intact in 1961 – rather than in the recovery of artefacts for sale or display – began in the 1950s. Scandinavians, French and Americans led the way with new equipment, notably the aqualung linked closely with the French pioneer Jacques Cousteau, making underwater work far easier. In England, the televised raising of the *Mary Rose* in 1982 brought marine archaeology to a mass audience. The university teaching of marine, or underwater, archaeology became more widespread from the 1980s taking its lead from the innovative work of the late Keith Muckelroy (Figs 12 and 13).

Currently, as defined broadly, underwater archaeology includes the study of shipwrecks and other vessels, including aircraft; structures built with underwater foundations such as crannogs and quays; and sites of human occupation that, for whatever reason, are now submerged. A practical division can be offered, however, between on the one hand the study of fixed structures such as buried land surfaces, and ports, and on the other formerly mobile objects such as ships and boats, with which this guide is concerned.

1.3 Understanding and quantifying the resource

As survey methods improve, and through chance discoveries, the number of historic ships and boats whose remains we are aware of will undoubtedly increase, probably markedly. In terms of quantifying the resource, three sets of figures are particularly useful: those contained within Historic England's National Record of the Historic Environment (NRHE), those resulting from a current initiative to identify hulk assemblages and those derived from an assessment of ships lost before 1840.

Records of ships and boats lost in English waters

In 1992, the National Monuments Record (now part of Historic England) began the systematic compilation of a record of historic shipwrecks and submerged archaeological sites within England's territorial waters. The record was created through desk-based research using both published and original archival research. This process of research and recording continues today and has been fully integrated with Historic England's National Record of the Historic Environment (NRHE), and provides a resource from which all wrecks in English waters can be assessed in their local, national, or international context (Fig 14).

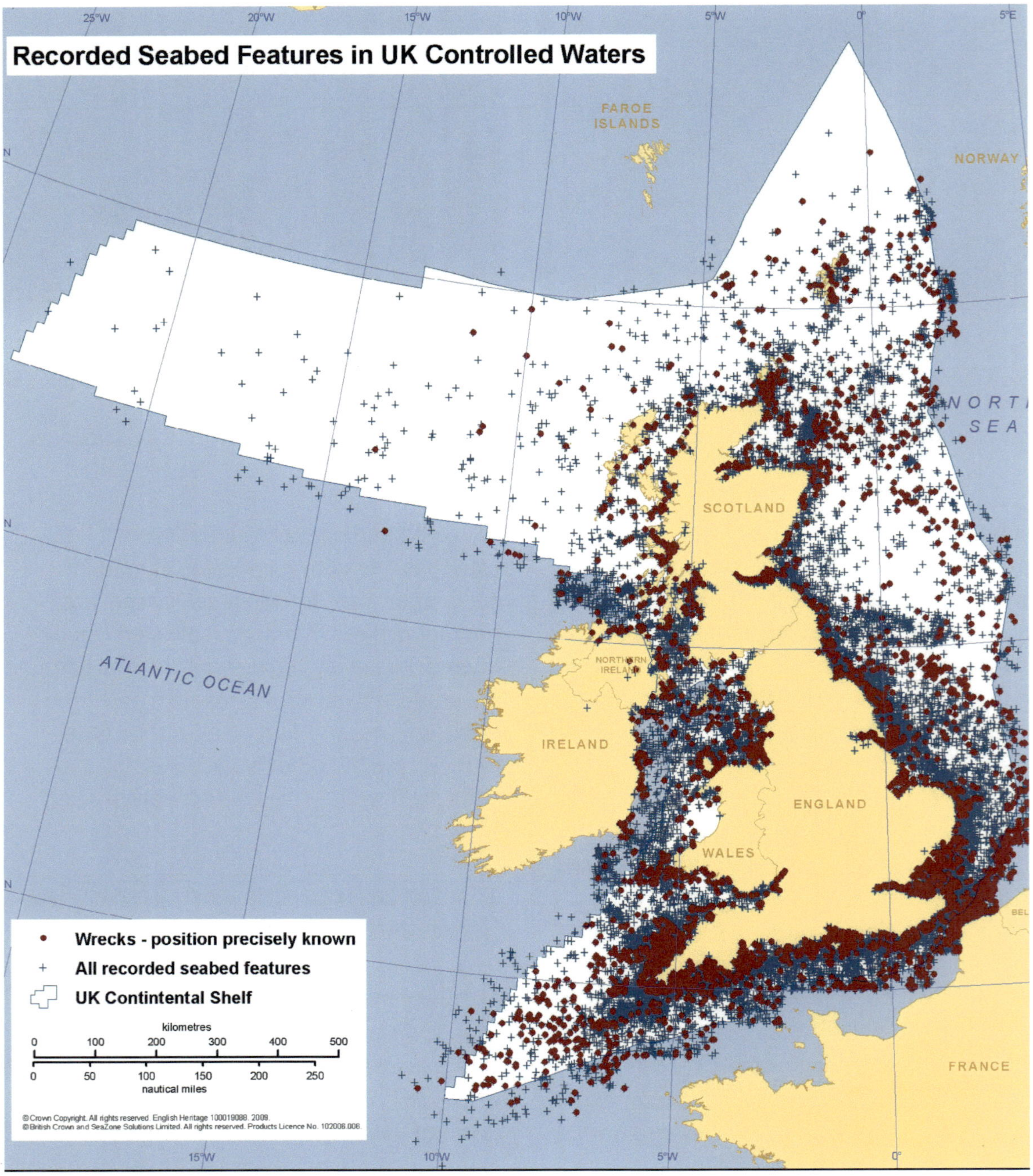

Figure 14
Recorded seabed features in UK waters.

The NRHE database of sites includes records of the following lost and wrecked vessels (including submarines) in English waters, the location and period of which is known:

Period	Number of sites
Pre-1499	40 (with an additional 73 logboats)
1500-1815	93
1815-1914	576
1914-1945	2216
1946-present	341

Hulk assemblages

Hulk assemblages comprise groups of vessels that have largely been stripped of fittings and are permanently grounded at the water's edge. This was often deliberate, to serve a specific purpose such as river bank reinforcement. Concentrations of hulks are particularly encountered in Gloucestershire, Cheshire and Dorset and along both shores of the Thames Estuary. To date, 188 hulk assemblages have been identified in England, many of them within areas of natural environment designation. Much work remains to be done, particularly with regard to recording the form and date of known groups in order to establish their importance.

Pre-1840 shipwreck sites

At present, very few ships and boats are offered statutory protection in England in comparison to the large numbers of known and dated wrecks and even greater numbers of recorded losses of vessels in English waters. Wreck sites that pre-date 1840 comprise just 4 percent of the 37,000 known and dated sites (as the majority of such sites are post-1914). One way of expanding the designation base has been to take a thematic approach. The theme of early ships and boats, from Prehistory to 1840, continues to be a strategic priority for Historic England because of the special technological, historical and human interest of these early vessels as well as their rarity in the archaeological record (Fig 15).

Figure 15
Reconstruction of the fifteenth-century Blackfriars Ship III, London.

2 Special Considerations

As noted above, the designation and recognition in other ways of ships, boats and underwater remains is a complex area, and one of the principal aims of this selection guide is to set out Historic England's current approaches, definitions and criteria. A primary concern will always be for the form of future management best suited to any particular vessel and, if *in situ*, its surrounds (Fig 16).

2.1 Listing

Listing has been sparingly deployed as a means of recognising the historic interest of some vessels. It falls into two broad categories, the first being the listing of intact vessels. Best known is the Grade I-listed *Cutty Sark*, permanently berthed in a dry dock in Greenwich, Greater London. Launched at Dumbarton in 1869, *Cutty Sark* was opened as a museum in 1957 and underwent a refit following a major fire in 2007. At Ocean Road, South Shields, South Tyneside, is found the Grade II-listed *Tyne*, the second-oldest lifeboat to survive, which dates from 1833 and which served for 60 years, saving over 1,000 lives; it stands within a cast-iron canopy of 1893 which is also listed.

Secondly, some listed buildings contain material taken from ships and boats, in general from vessels broken up at the end of their service, less commonly from wreck sites. Perhaps best-known is the Grade II*-listed Chesapeake Mill at Wickham (Hampshire), partly constructed from timbers salvaged from the USS *Chesapeake*, captured in 1813 after a frigate action against HMS *Shannon* and subsequently broken up. Recent investigation of the mill's fabric has enabled light to be shed on the original construction of the *Chesapeake*: a rare documented example of the widely-held belief that ships' timbers were extensively used in domestic construction. Surviving at Shotley (Suffolk) is a Grade II-listed mast, 44 metres high, serving the former training establishment HMS *Ganges*: the lower part came from HMS *Cordelia* (launched 1881) while the upper part came from HMS *Agincourt* (launched 1865).

Numerous listed churches contain maritime elements: the altar rails of St John the Baptist church, Pewsey (Wiltshire) are made from timber taken from the 112-gun *San Josef*, boarded by Nelson at Cape St Vincent in 1797, for instance; and the pre-Reformation rood screen at the Grade I-listed Gunwalloe church (Cornwall) is said to have been constructed from salvaged timber from the *St Anthony of Lisbon*, which foundered nearby in 1527. Secular buildings re-used nautical material too: the Grade II-listed White Swan Hotel at Alnwick (Northumberland) includes a room lined in Louis XV-style panelling from the SS *Olympic*, sister ship to the *Titanic*, which was broken up at Jarrow in 1935. Figureheads included within larger listings include that of the *Roseau*, a Jersey-built barque of 1857 now installed within

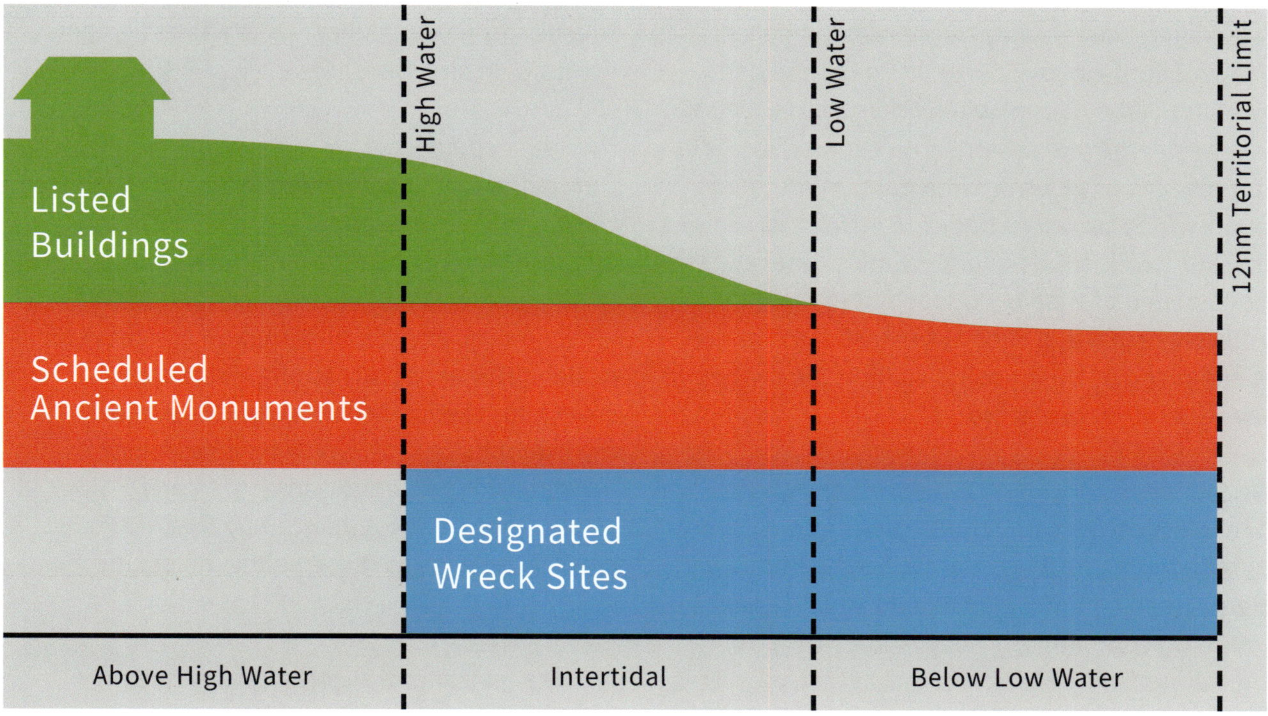

Figure 16
Extent of statutory designations in England.

the Grade II-listed Cruise Terminal 1 at Dover(Kent) of 1914. In cases such as these, the special interest afforded by the maritime elements can play a crucial part in determining whether listing is warranted. For more information about our approach to listing in these cases, see our Maritime and Naval Buildings Selection Guide.

Overall, listing will be Historic England's preferred designation response when a vessel has effectively entered the planning system as a permanently grounded structure which has been fixed to dry land, or whose components have been deliberately incorporated into a permanent terrestrial structure - for instance, a dry dock. Listing will seldom, if ever, be appropriate for vessels still capable of being moved to new moorings or berths, or for abandoned vessels (or their remains) which have come to rest on the ground (say a river bank or mud flat). In some cases, particularly where restoration, maintenance and replacement are ongoing, other permanently moored vessels of high significance have not been listed: these include HMS *Victory* in Portsmouth and the SS *Great Britain* in Bristol.

2.2 Scheduling

Although the 1979 Ancient Monuments and Archaeological Areas Act makes provision for the designation of vessels or the remains of vessels, in practice the Act has rarely been used to designate such sites in England. As with listing, in order for a vessel to be considered for scheduling, it must be a permanently grounded structure, preferably within an archaeological context such as sunk in a riverbank, and for those in marine contexts sited within English Territorial waters (that is out to 12 nautical miles).

Isolated examples of vessels are included in the Schedule, for example the rare second-century AD Romano-British barge at Bermondsey (London Borough of Southwark). A more recent addition to the Schedule, a Second World War Phoenix Caisson (part of a 'Mulberry' harbour, constructed as part of the D-Day Allied invasion of Normandy in 1944) located on the coast of Kent at Littlestone-on-Sea, was given protection on account of innovative engineering employed in its construction and rarity. Such designations are likely to remain the exception.

More often, vessels may be included as part of the scheduling of a wider site. The scheduled Waltham Abbey Gunpowder works, active for 300 years from the mid seventeenth century, includes the remains of seven powder barges and three punts which served to transport raw materials and finished products around the work's canal system. Constructed of timber, copper and leather, the vessels are unusual survivals whose significance is enhanced by their position within the context of the wider Gunpowder works.

Where vessels are identified during the course of work consented under planning regimes, it may be more effective to preserve their heritage significance by managing it in accordance with the National Planning Policy Framework or the UK Marine Policy Statement through mechanisms such as preservation *in situ*, or by recording and publication, or by recovery of the vessel and its relocation to a museum context.

Selection criteria

The particular considerations used by the Secretary of State when determining whether sites of all types are suitable for statutory designation through scheduling are set out in www.gov.uk/government/publications/scheduled-monuments-policy-statement. These considerations are under the following headings, usually termed 'non-statutory criteria': period; rarity; documentation/finds; group value; survival/condition; fragility/vulnerability; diversity; and potential.

Period

Vessels from all periods are important in reflecting technological advances in construction and materials, and provide evidence of trade networks, industry, and transport. Those vessels which best illustrate or epitomise this development have a strong claims to national importance.

Rarity

The rarity of vessels' remains for periods before 1700 is such that any firmly dated vessels from this period are likely to be of national importance and may merit scheduling. For vessels of later date, particularly those types for which examples survive today, scheduling will always be under exceptional circumstances only.

Documentation/Finds

Our understanding of shipbuilding, transport, trade and industry can be greatly enhanced by the survival of historical documentation relating to particular vessels and their service. Where interpretive documentation can provide evidence for especially strong historical claims, for example confirming a ship to be the last of its type, this may be a key factor in establishing its importance. Similarly, significance can be enhanced by the existence of artefacts such as those held in museums.

Group Value

In some instances a vessel's importance may be strengthened by an association with other vessels of a similar type, for example the group of gunpowder boats at Waltham Abbey (Essex), which allow for comparative study. Association within a wider context which reflects their use can also be a consideration. In the case of hulks (vessels that have been stripped and abandoned), as well as having intrinsic interest, they can contribute to the story of a landscape, and its long-term evolution and management.

Survival/Condition

Given the range of materials used in boat-building, survival of vessels can be highly varied, from the survival of the sand-imprint of the ship at Sutton Hoo or fragment of the log boat at Shardlow (Derbyshire) to the concrete boats of Second World War date at Purton (Gloucestershire). Given the rarity of surviving vessels of pre-1700 date, even fragmentary survivals are likely to be of national importance although a judgment must be reached as to the degree of survival and intactness. For vessels of later date, increasingly complete survival, allied to strong archaeological and historical importance, will be expected before scheduling would be considered.

Fragility / Vulnerability

Highly important archaeological evidence from some wrecks can be destroyed by the selective or uncontrolled removal of material by unsympathetic treatment by works or development or by natural processes. Some vessel types are likely to be more fragile than others and the presence of commercially valuable objects within a wreck may make it particularly vulnerable. Vulnerable sites of this type would particularly benefit from protective designation.

Diversity

The policy statement cited above notes that assets may be selected for designation because they possess a diverse combination of high quality features, others because of a single important attribute. The importance of wrecked vessels can reflect the interest in their architectural design, decoration and craftsmanship, or their technological innovation or virtuosity, as well as their representativeness. Consideration should be given both to the diversity of forms in which a particular vessel type may survive and to the diversity of surviving features. Some vessels types may be represented in the surviving record by a wide variety of building types and techniques which may be chronologically, regionally, or culturally conditioned. The sample of protected sites should reflect this wide variety of forms. In addition, some wrecks may be identified as being of importance because they possess a combination of high quality surviving features or, occasionally, because they preserve a single important attribute.

Potential

England's maritime past is one of its most defining characteristics throughout all periods. Evidence for the construction and use of vessels gives us great insight into not only the exploitation of our immediate marine environment, but also into the development of wider trade and transport networks. This is especially true of earlier periods which are lacking in the rich literature and documentation of later times. Surviving vessels may also provide evidence of their use and construction, reflecting technological developments which in some instances may be all but lost. For the prehistoric period, in particular, the remains of vessels may be some of the largest artefacts discovered which demonstrate the technology of woodworking and management of woodland resources. Similarly, where vessels are found ***in situ***, associated deposits may be rich in palaeo-environmental remains. The potential which a vessel has for answering questions about our maritime past will be a consideration in establishing its importance. If remains of a cargo survive it is likely to add very considerably to the vessel's significance, for its evidence of trade and material culture at a particular point in time.

The 1979 Act also allows for a site to be designated on account of its historic interest. This criterion can be split into considerations of historical associations, and those relating to social, economic and mercantile history. For the purposes of this guide, historical association is taken to mean any identified vessel which played a key role in England's national history or which is associated with a historical figure of national significance. Where this can be firmly established historic significance will be a compelling factor in assessing the vessel's national importance. Historical claims can be particularly relevant for vessels in terms of their cargos, their crews, their construction, the trades they plied and the routes they served.

Finally, the Act also allows for a site to be scheduled for its artistic interest. Carved decoration or a notable figurehead may enhance a vessel's claims to national importance, as it may if a cargo of considerable artistic interest is known to survive.

Vehicles, aircraft and structures lost in maritime contexts

Vessels are not the only forms of mobile object to be found in a marine setting. The Ancient Monuments and Archaeological Areas Act 1979 allows for the scheduling of submerged assets including the remains of any vehicle, aircraft or other moveable structure or lost section thereof. As with shipping, many such assets represent losses which occurred during the Second World War. Relatively few such items have been designated to date, although on the West Knock sandbank off Southend (Essex) is a scheduled 'Phoenix' caisson, a section of temporary 'Mulberry Harbour' intended to be used in the Normandy landings following D-Day for the rapid re-supply of troops but which sank en route from Immingham (Lincolnshire); a comparable example in Poole Harbour (Dorset) has been listed. Examples of vehicles on the sea-bed – none of which are scheduled – include roughly ten amphibious tanks lost in training for D-Day, while a Dornier 17 German bomber, shot down in 1940, was found on the Goodwin Sands off Kent in 2011. In this latter instance, this is the only known example of this type of aircraft: prior to its recovery in 2013, questions were asked as to whether *in situ* preservation was the more desirable option. See also below (Section 3) for the special provisions which apply in respect of military aircraft and military maritime graves.

2.3 Protected Wreck Sites

The Protection of Wrecks Act (1973) allows the Secretary of State to designate a restricted area around a wreck to prevent uncontrolled interference. Designated sites are identified as being likely to contain the remains of a vessel, or its contents, which are of historical, artistic or archaeological importance. Access is monitored through a system of licensing administered through Historic England on behalf of the Secretary of State. The criteria are the same as for scheduling (see above) (Fig 17).

Each restricted area comprises an exclusion zone and it is an offence to tamper with, damage or remove any objects or part of any vessel, or to carry out any unauthorized activities within this restricted area. Administration of the 1973 Act, and the associated licensing scheme, is the responsibility of Historic England in England on behalf of the Secretary of State (Fig 18).

Figures 17 (top) and 18 (bottom)
Top: West Bay Protected Wreck Site (Dorset). Seventeenth-/eighteenth- century bronze cannon and iron cargo.

Bottom: Hoisting 350-year-old gun carriage from the protected wreck of the *London* from the crane barge to quayside at Leigh-on-sea, Essex.

3 Where to Get Advice

Other Means of Recognising and Protecting Historic Vessels

3.1 The National Register of Historic Vessels

National Historic Ships UK, with its governing Council of Experts, was established in 2006 as a non-departmental public body reporting to the Department of Culture, Media and Sport with a specific remit to advise the Secretary of State and other public bodies on ship preservation and funding priorities. Since 2006, it has maintained a non-statutory National Register of Historic Vessels, which now includes over 1,000 vessels. Registration provides an authoritative assessment of vessels' significance, but places no restrictions on how vessels are maintained, repaired or used, and offers no formal protection (Fig 19).

So far as possible, Historic England works with this body to ensure appropriate recognition that avoids overlap.

Figure 19
HMS ***Warrior,*** Portsmouth, launched 1860 and now part of the National Historic Fleet.

3.2 Military maritime graves

The Protection of Military Remains Act 1986, administered by the Ministry of Defence, applies both to the remains of service personnel and also to the vessels and aircraft in which they were lost. It secures them from unauthorized interference and is thus an important means of protection. Remains fall under two headings: protected places and controlled sites. Wrecks are designated by name and can be designated as *protected places* even if the location of the site is not known. Thus, the wreckage of a UK military aircraft is automatically a protected place even if the physical remains have not been previously discovered or identified. Shipwrecks need to be specifically designated, and designation as a protected place applies only to vessels that sank after 4 August 1914. The Act makes it an offence to interfere with a protected place, to disturb the site or to remove anything from the site. Divers may visit the site but the rule is look, don't touch and don't penetrate the hull of a protected vessel. *Controlled sites* containing the remains of an aircraft or a vessel that crashed, sank or was stranded within the last two hundred years must be specifically designated by location. The Act makes it illegal to conduct any operations (including any diving or excavation) within the controlled site that might disturb the remains unless licensed to do so by the Ministry of Defence. Controlled sites at sea are marked on Admiralty charts and their physical location may be marked by means of a buoy.

For those sites covered by the Military Remains Act, to avoid superfluous protection, Historic England's view is that statutory heritage designation will seldom be appropriate.

3.3 Designation of the natural environment

There are a number of national environmental designations which place constraints over activities on riverine, coastal and marine locations where ships and boats, or their remains, may be found. Historic England suggests that using the protection afforded by such designations to uphold a site's historic or cultural value may also be an appropriate designation response. Environmental designations include[a] Sites of Special Scientific Interest (administered by Natural England), defined as England's very best wildlife and geological sites, and [b] Ramsar Sites (advised by the UK statutory nature conservation agencies), being wetlands of international importance, designated under the Ramsar Convention, an inter-governmental treaty, signed in 1971.

The Marine and Coastal Access Act 2009 (Part 5) also enables Ministers to designate and protect Marine Conservation Zones (MCZs). Lundy Island in the Bristol Channel, a former Marine Nature Reserve, became the first MCZ in January 2010.

3.4 Contact Historic England

East Midlands
2nd Floor, Windsor House
Cliftonville
Northampton NN1 5BE
Tel: 01604 735460
Email: eastmidlands@HistoricEngland.org.uk

East of England
Brooklands
24 Brooklands Avenue
Cambridge CB2 8BU
Tel: 01223 582749
Email: eastofengland@HistoricEngland.org.uk

Fort Cumberland
Fort Cumberland Road
Eastney
Portsmouth PO4 9LD
Tel: 023 9285 6704
Email: fort.cumberland@HistoricEngland.org.uk

London
1 Waterhouse Square
138-142 Holborn
London EC1N 2ST
Tel: 020 7973 3700
Email: london@HistoricEngland.org.uk

North East
Bessie Surtees House
41-44 Sandhill
Newcastle Upon Tyne
NE1 3JF
Tel: 0191 269 1255
Email: northeast@HistoricEngland.org.uk

North West
3rd Floor, Canada House
3 Chepstow Street
Manchester M1 5FW
Tel: 0161 242 1416
Email: northwest@HistoricEngland.org.uk

South East
Eastgate Court
195-205 High Street
Guildford GU1 3EH
Tel: 01483 252020
Email: southeast@HistoricEngland.org.uk

South West
29 Queen Square
Bristol BS1 4ND
Tel: 0117 975 1308
Email: southwest@HistoricEngland.org.uk

Swindon
The Engine House
Fire Fly Avenue
Swindon SN2 2EH
Tel: 01793 445050
Email: swindon@HistoricEngland.org.uk

West Midlands
The Axis
10 Holliday Street
Birmingham B1 1TG
Tel: 0121 625 6870
Email: westmidlands@HistoricEngland.org.uk

Yorkshire
37 Tanner Row
York YO1 6WP
Tel: 01904 601948
Email: yorkshire@HistoricEngland.org.uk

4 Select Bibliography

4.1 General

Brown, P., *Britain's Historic Ships: The Ships That Shaped the Nation: A Complete Guide* (2009)

Cant, S., *England's Shipwreck Heritage: From Log-boats to U-boats* **(2011)**

Catsambis, A., Ford, B. and Hamilton L.H., *The Oxford Handbook of Maritime Archaeology* (2011)

Dromgoole, S., *Underwater Cultural Heritage and International Law* (2013)

Dunkley, M., 'The Value of Historic Wreck Sites', pages 18-28 in Radic Rossi, I., Gaspari, A. and Pydyn, A. (eds.), *Proceedings of the 13th Annual Meeting of the European Association of Archaeologists* (2008)

Larn, R. and Larn, B., *Shipwreck Index of the British Isles*, 5 vols. (1995-2000)

Marsden, P., *Ships and Shipwrecks* (1997)

Ransley, J. and Sturt, F, *People and the Sea: A Maritime Archaeological Research Agenda for England* (2013)

4.2 Prehistoric and Roman

Dean, M., 'Evidence for Possible Prehistoric and Roman Wrecks in British Waters' *International Journal of Nautical Archaeology* **13.1** (1984), 78-80

Gardiner, R., *The Earliest Ships: The Evolution of Boats into Ships* (2004)

Marsden, P., *Ships of the Port of London: First to Eleventh Centuries AD* (1994)

McGrail, S., 'Boats and Boatmanship in the Late Prehistoric Southern North Sea and Channel Region', pages 32-5 in McGrail, S. (ed.), *Maritime Celts, Frisians and Saxons* (1990)

McGrail, S., *Ancient Boats in North-West Europe: The Archaeology of Water Transport to AD 1500* (1998)

Roberts, O. T. P., 'Interpretations of Prehistoric Boat Remains' *International Journal of Nautical Archaeology* **35.1** (2006), 72-8

4.3 Medieval

Flatman, J., *Ships and Shipping in Medieval Manuscripts* (2009)

Friel, I.,*The Good Ship. Ships, Shipbuilding and Technology in England 1200-1520* (1995)

Hutchinson, G., *Medieval Ships and Shipping* (1994)

Marsden, P., *Ships of the Port of London: Twelfth to Seventeenth Centuries AD* (1996)

Rodger, N.A.M., *The Safeguard of the Sea: A Naval History of Britain Volume 1, 660-1649* (1997)

4.4 Post-Medieval

Farquharson-Roberts, M., *A History of the Royal Navy: World War One* (2014)

Padfield, P., *Rule Britannia: the Victorian and Edwardian Navy* (2002)

Redford, D., *A History of the Royal Navy: World War Two* (2014)

Rodger, N.A.M., *The Wooden World: An Anatomy of the Georgian Navy* (1986)

Rodger, N.A.M., *The Command of the Ocean: A Naval History of Britain Volume 2, 1649-1815* (2004)

4.5 Websites

https://www.gov.uk/government/policies/conservation-of-historic-buildings-and-monuments
Department for Culture, Media and Sport web pages on the conservation of historic buildings and monuments. [Site accessed: July 2016]

https://www.gov.uk/government/publications/protection-and-management-of-historic-military-wrecks-outside-uk-territorial-waters
Department for Culture, Media and Sport & Ministry of Defence joint guidance on how policies and legislation apply to historic military wreck sites which are discovered outside UK territorial waters. [Site accessed: July 2016]

http://nationalhistoricships.org.uk
National Historic Ships UK. The website includes databases on Registered Historic Ships and the National Historic Fleet and information about the selection process. [Site accessed: September 2015]

http://thesaurus.historicengland.org.uk/
National Monuments Record Thesaurus (for Maritime Craft Types).[Site accessed: July 2016]

http://archaeologydataservice.ac.uk/
Online project archives for selecting marine sites that are sufficiently archaeologically important to warrant special measures, particularly the *Selection Guide: Boats & Ships in Archaeological Contexts*. [Site accessed: July 2016]

5 Acknowledgements

Images

© Historic England

Front cover and Figures 1, 2, 3, 8, 9, 12 and 15

© Wessex Archaeology

Figures 7, 10, 13, 14, 16 and 17

© Others

Figure 4: Borough of Pools Museum Service

Figure 5: geocities.com/botleyhistory

Figure 6: Paul Stamper

Figure 11: Image courtesy of the Channel Coastal Observatory

Figure 18: Museum of London

Every effort has been made to trace the copyright holders and we apologise in advance for any unintentional omissions, which we would be pleased to correct in any subsequent editions.

This page is left blank intentionally

Historic England

We are the public body that looks after England's historic environment. We champion historic places, helping people understand, value and care for them.

Please contact guidance@HistoricEngland.org.uk with any questions about this document.

HistoricEngland.org.uk

If you would like this document in a different format, please contact our customer services department on:

Tel: 0370 333 0607
Fax: 01793 414926
Textphone: 0800 015 0174
Email: customers@HistoricEngland.org.uk

All information and weblinks accurate at the time of publication.

Please consider the environment before printing this document

HEAG143
Publication date: May 2012
Reissue date: September 2016
Design: Historic England

Made in the USA
Monee, IL
05 November 2026

67987966R00019